# QUANTUM PHYSICS FOR BEGINNERS

DISCOVER HOW THE QUANTUM PHYSICS PHENOMENA INFLUENCE YOUR WORLD IN A EASY AND INTUITIVE WAY WITH NO HARD MATH.

**EUGENE SMITH**

**Copyright - 2020 - All rights reserved.**

The content contained within this book may not be reproduced, duplicated or transmitted without direct written permission from the author or the publisher.

Under no circumstances will any blame or legal responsibility be held against the publisher, or author, for any damages, reparation, or monetary loss due to the information contained within this book. Either directly or indirectly.

**Legal Notice:**

This book is copyright protected. This book is only for personal use. You cannot amend, distribute, sell, use, quote or paraphrase any part, or the content within this book, without the consent of the author or publisher.

**Disclaimer Notice:**

Please note the information contained within this document is for educational and entertainment purposes only. All effort has been executed to present accurate, up to date, and reliable, complete information. No warranties of any kind are declared or implied. Readers acknowledge that the author is not engaging in the rendering of legal, financial, medical or professional advice. The content within this book has been derived from various sources. Please consult a licensed professional before attempting any techniques outlined in this book.

By reading this document, the reader agrees that under no circumstances is the author responsible for any losses, direct or indirect, which are incurred as a result of the use of information contained within this document, including, but not limited to, - errors, omissions, or inaccuracies.

# Table of Contents

- **INTRODUCTION** 5
- **CHAPTER 1**
  What is Quantum Physics? 9
- **CHAPTER 2**
  Quantum Physics - The Localization of Manifestation 23
- **CHAPTER 3**
  Quantum Theory - An Overview of the Mystifying Science 27
- **CHAPTER 4**
  Quantum Physics and You 37
- **CHAPTER 5**
  The Building Blocks of Matter and Wave-Particle Duality 49
- **CHAPTER 6**
  Quantum Possibilities and Waves 61
- **CHAPTER 7**
  Application: Quantum Computing 67
- **CHAPTER 8**
  Why are Quantum Computers so Difficult to Make? 83
- **CONCLUSION** 93

# Introduction

Quantum physics, like every other science, is an attempt to define reality. Although quantum physics is by definition a science, its surprising results have blurred the line between classical science and its centuries-old rivals of philosophy, religion, and mysticism.

As it occurred, the quantum line of inquiry has not established truth, but rather redefined what reality is. It's as if we were trying to tune up the car, and somewhere along the way we figured out that the car wasn't a car, but instead a flubbermobile. Deepak Chopra said, "Not only is the Universe stranger than we think, it is stranger than we can think." This is definitely a true statement when one tries to understand quantum physics.

But despite this, the fundamental ideas behind quantum physics make a lot of sense when you keep them to an intuitive level. Consider the following three basic principles:

1. The core tenet of quantum physics is that there is a wavelike ocean of possibilities at the very base level of matter, which physicists have called the "unified field." Matter and form, all arise from this undefined, nebulous

network of potentials.

2. Experiments with subatomic particles have shown that physical space and time are irrelevant at this almost inconceivable level of matter. Not only can the particles present themselves as waves or particles, they can also exist at the same time in two places (called "superposition"). Quantum science has been able to also produce evidence that not only does past experience affect the present (a finding that we have come to expect), but that future events can also bring about changes in present reality.

3. Linked to the "construction of the mind designed to create form and order in the face of a chaotic universe." As early as 1927, physicists learned about the importance—and impact—of the observer in any given reality. The famous Heisenberg Uncertainty Principle arose from the findings that not only does a particle (reality) change when it is observed, that if/when an observer changes how they think/perceive a certain reality, that reality changes along with it.

So, there is no such thing as absolute, objective reality. It's all subjective.

What would it mean if that statement were true? Some people may raise their hands and say, "Let anarchy begin," or "This idea will foster a culture of egomaniacs," or "How, then, do we think of universal realities such as Nature, Justice, and Truth?"

To which one may respond, "Chaos reigns without us granting him permission," or "Take a look around; haven't we already created a society of egomaniacs?" or "The world is constantly at war because of the irreconcilably subjective interpretations of Religion, Justice, and Reality."

When you pause and consider the question honestly, it becomes more and more evident that our perceptions of reality are already subjective. Simple proof rests in the fact that, if twenty-five of us had gone to the same group, there would have been several different experiences of the same thing.

Irrespective of that, the idea that anything and everything is subjective is rather controversial. No surprise then, that eighty years after the first "quantum" discoveries, we still find ourselves in the throes of a paradigm that focuses on one clear and utterly valid truth. Unfortunately, there is no space for imagination, consciousness, faith, or harmony in this almost universally acknowledged "clockwork universe." And this is a very nice place to sit down for those who neglect (or, worse, mistrust) imagination, morality, faith, or unity.

Fortunately, as a result of the discoveries of quantum physics, we will all have to learn to think differently. The dawn of the 21st century provides the opportunity for a new heaven, a new planet, and to use the term that has already been used, a new paradigm for understanding our universe. We truly have a new world at our doorstep.

Since you are fascinated by the world of quantum physics, I strongly suggest you pay close attention to every theory and law relatively explained in this book as we journey to the quantum physics world.

# Chapter 1

## What is Quantum Physics?

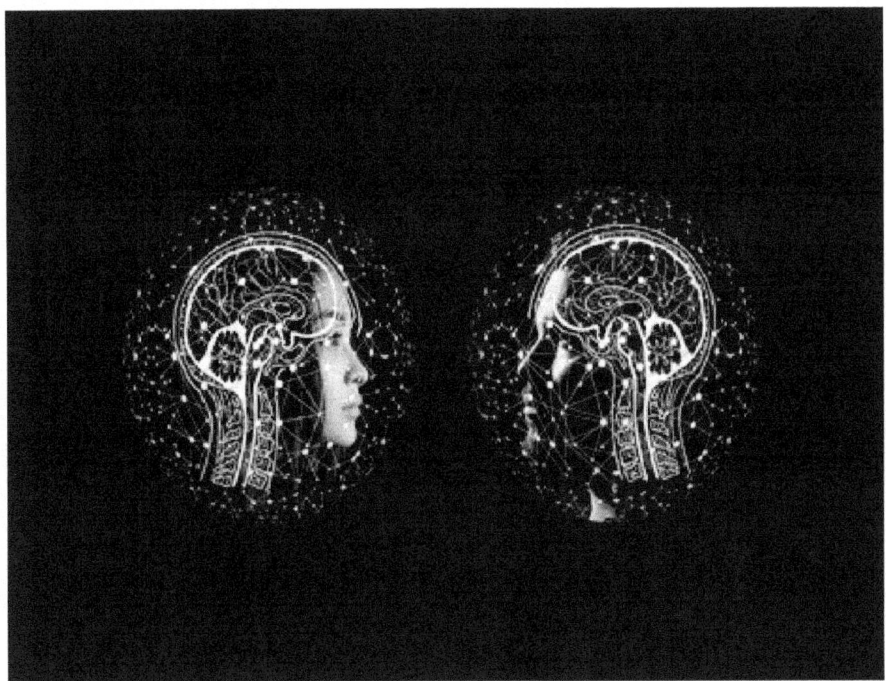

Quantum Mechanics, commonly referred to as Quantum Physics, is the interaction between energy and matter. The term "quantum" is Latin for "how many." The mechanics of this refers to a unit that quantum theory assigns to certain physical quantities in very small quantities as a measure. In essence, quantum expressions are usually examined and studied at a sub-atomic level with sub-atomic particles.

Sub-atomic particles are tiny. If the atom were as large as the room, the sub-atomic particle would be as large as the gum drop within the kitchen cupboard of that building. There are a few things that had to happen before the theory of Quantum Mechanics took root. In 1838, after the discovery of cathode rays, Gustav Kirchoff, in 1850, published a statement on the "black body radiation" issue. Then, in 1877, Ludwig Boltzmann proposed that the energy states of the physical system could be discreet.

In 1900, Max Planck came up with a theory that energy was radiated and absorbed, and then he produced a formula that would be known as "Planck 's Action Constant."

Planck is also regarded as Quantum Mechanics' patriarch. After his theory was written, other scientists took notice of it, and immediately you had a few more theories to build before Quantum Mechanics was theorized and studied all over the world.

It's because of Quantum Mechanics that we're on the brink of anti-gravity, have super conductors, MRI devices in hospitals, and now we're also able to imagine that time travel is possible.

This all sounds so amazing, but this is what the scientists working in the world of quantum mechanics are going to tell you. The most difficult thing for most of us to understand is the interaction between sub-atomic particles and the Law of Attraction.

In the Quantum Mechanics test, it was found that sub-atomic particles are moving in a direction. Any other force is moving the building blocks of physical matter across the universe.

After a few double-blind tests, using 'Sub-Atomic Particles' as subjects, it was discovered that they could change from particles to wave form and then back again. These particles could leave our dimension and pop right back into it again. We also found that these sub-atomic particles, depending on the intent, have changed from particles to waveforms. Then we discovered that we weren't able to get rid of the equation while testing the particles. We've affected the particles by worrying about the result. There's a lot more to it.

It's all becoming very complicated. Einstein remained uncertain until the day he died. Understanding particle/wave duality is not something that most of us can comfortably wrap our minds around.

Yet one of the ideas that has developed from the base of Quantum Physics is that we influence the very fabric of life by thinking about it. Our minds have a phrase that goes out, and then returns to us what we're focused on. It's the Law of Attraction.

A moving clock is running slowly as its speed increases and starts running at the speed of light, unless you travel at the same pace as the clock, in which case it appears to be behaving as usual. As the speed of an object increases, its apparent size decreases and its weight increases until the speed of light is reached, when it disappears and its weight is infinite. Luckily, none of this is really detectable until you get closer to the speed of light. In the same way, quantum physics has shown that, although the world at large may seem to work logically and predictably at the subatomic level (dealing with whatever it is that makes up the atoms), things are very different. For instance:

- You can prove that sub-atomic entities are both matter and energy-it just depends on the experiments you do. There is therefore nothing as a detached observer. What you see is tied up with preconceptions of your own.

- You can't guess what a sub-atomic object is going to do — you can just say what it's going to do. Thus, the universe can no longer be seen as an inanimate machine following logical rules.

- Entities from a common source remain connected even if they are separate, so that what happens to one is immediately reflected in the other. In this case, relationships between entities are more important than their separate identities, and the entire universe can be interconnected.

In an attempt to make sense of all this, the physicist, David Bohm, suggests that reality is not just made up of matter and energy, but of matter (explained order), energy (implicit order), and meaning (super-implicit order). Each contains the other two-matter, consisting of energy and meaning, energy of thirds and meaning, and the sense of matter and energy. Truth, therefore, is not two-dimensional as an image, but multidimensional as a hologram, with each component capable of representing the whole and requiring the whole to make sense.

Some people use quantum physics to "prove" the validity of what mystics and magicians have been saying for millennia-but do they really need to prove it? Moreover, a lot of quantum physics is theoretical. What has really been done is to narrow the gap between science and religion. Buddhism is often defined as a road to enlightenment for people of intellectual temperament. The same may be said for modern physics. The rest of us are going to muddle on in our normal way, and-who knows? We will all definitely end up in the same place.

Quantum science is the foundation of how atoms work, and why chemistry and biology work as they do. You, me and the gatepost – at least at some level, we're all dancing to the quantum tune. If you want to understand how electrons travel through a computer chip, how photons of light transform into electrical current in a solar panel, or amplify themselves in a laser, or even how the sun burns, you'll need to use quantum physics.

The complexity and, for the physicists, the fun begins here. First of all, there is no single quantum theory. There is quantum mechanics, the fundamental mathematical structure that underpins all that was first introduced in the 1920s by Niels Bohr, Werner Heisenberg, Erwin Schrödinger and others. It characterizes simple things like how the position or momentum of a single particle or a group of few particles changes over time.

But to completely understand how things work in the real world, quantum mechanics must be coupled with other elements of physics — primarily, Albert Einstein's special relativity theory, which describes what happens as objects move very quickly — to establish what is known as quantum field theories.

Three different quantum field theories deal with three of the four fundamental forces that matter interacts with: electromagnetism, which explains how atoms hold together; strong nuclear power, which explains the stability of the nucleus at the heart of the atom; and weak nuclear power, which explains why some atoms are subject to radioactive decay.

For the last five decades or so, these three theories have been put together in a ramshackle coalition known as the "normal model" of particle physics. For all the fact that this model is slightly kept together with sticky tape, it is the most precise image of the basic work of the matter that has ever been invented. The crowning glory came in 2012 with the discovery of the Higgs boson, a particle that gives mass to all other fundamental particles, the existence of which had been predicted on the basis of quantum field theories as far back as 1964.

Conventional quantum field theories work well in explaining the effects of experiments with high-energy particle crushers, such as CERN's Large Hadron Collider, where the Higgs have been found, whose probe is the smallest in size. But if you try to understand how things work in a much less abstract circumstance – how electrons travel or don't travel through a solid material to make a substance, a metal, an insulator or a semiconductor, for example – things get much more complicated.

Billions and billions of experiences in such crowded environments call for the creation of "successful field theories" that gloss over some of the gory information. The challenge in building such theories is why many important questions in solid-state physics remain unanswered – for example, why at low temperatures certain materials are superconductors that require current without electrical resistance, and why we can't get this trick to work at room temperature.

But there is a massive quantum mystery underneath all these practical problems. At the basic level, quantum physics predicts very strange things about how matter works, which are completely at odds with how things seem to work in the real world. Quantum particles can act as particles, located in a single place; or they can act as waves, scattered throughout space or in several places at once. How they seem to depend on how we want to measure them, and before we measure them, they seem to have no definite properties at all – leading us to a profound conundrum about the existence of basic truth.

This fuzziness leads to apparent paradoxes, like the Schrödinger's cat, in which a cat is left dead and alive at the same time, thanks to an uncertain quantum process.

But that's not all of it. Quantum particles also seem to affect each other instantaneously, even if they are far away from each other. This genuinely bamboo-like phenomenon is known as entanglement, or, in a phrase coined by Einstein (a great critic of quantum theory), "spooky behavior at a distance." These quantum powers are completely alien to us, but they form the basis of emerging technologies such as ultra-secure quantum cryptography and ultra-powerful quantum computing.

But no one knows what it all entails. Some people think we just have to agree that quantum physics describes the material world to the degree that we find it impossible to fit our experience in the broader, "classical" world. Some think there must be a stronger, more intuitive explanation out there that we have yet to discover.

There are several elephants in the room in all of this. For example, there is a fourth basic force of nature that has so far been unable to describe quantum theory. Gravity remains the territory of Einstein's general Theory of Relativity, a firmly non-quantum theory that does not even involve particles. Intensive efforts over decades bring gravity under the quantum umbrella and thus explain all fundamental physics within one "theory of all" have come to nothing.

In the meantime, cosmological measurements indicate that more than 95% of the universe consists of dark matter and dark energy, things for which we actually have no explanation within the standard model, and conundrums such as the nature of the role of quantum physics in the chaotic workings of existence remain unknown.

Why are some objects well described by classical physics models, while others require a description of quantum physics?

There are two main reasons for this: smallness and coherence, each of which is briefly summarized here. Smallness can refer to different aspects of objects: smallness of size or small energy content. If the object is roughly the size of an atom (about 10 metres), then it can almost certainly not be accurately modeled using classical mechanics, and must be described in a more accurate quantum theory. Interestingly, however, the opposite is not necessarily true; objects as large as a millimeter (about a twentieth of an inch) have been observed in experiments displaying behaviors that indicate a quantum nature.

For example, the smallness (or lowness) of the energy content could refer to a tiny electrical current in a metal wire (a superconductor) at a temperature slightly higher than the absolute zero (- 273 degrees Celsius or - 459 degrees Fahrenheit). Low temperature means low energy. Or it could apply to a brief flash of light containing only a small fraction (say, 10-21) of the energy emitted by a one-hundred-watt bulb in one second. Such a flash of light is said to contain only one 'photon' of light, which refers to the smallest discreet amount of energy that the light of a certain color can carry. A discreet amount of energy such as this is also called a 'light quantum.'

The plural quantum is 'quantum.' For example, a burst of light with a large amount of energy is said to contain many quantums. This discreetness of the energy carried in light, which we will explore in more detail later, is the origin of the name 'quantum physics.'

In principle, a single quantum entity, such as a photon, could be extended over a very large volume—many kilometers, for example. Although such a photon would be large in size, it would be very small or low in energy content, and therefore the quantum theory would still apply to it. The second general reason that an object can require a quantum definition is 'quantum coherence.' Quantum coherence is a subtle concept and cannot be properly understood until one understands how the status of an object is represented using quantum theory. In the case of an electron, quantum coherence enters the theory of how it accounts for the different possibilities that may exist before the location of the electron is observed. In a way, the standard laws of logical thinking, such as stating, "It is located here or not located here," do not apply to quantum objects. Instead, it says, "Both possibilities must be superimposed in our thinking and not considered separately."

## How Was Quantum Physics Discovered?

The thorough story of how quantum physics was discovered is interesting, and a lot of books say it. But with hindsight, it seems to me that the great struggle to invent quantum theory in the early twentieth century, says more about the difficulty that humans had (and still have) in going beyond classical physics thinking than the facts of quantum physics say. So this book is not about the historical aspects of physics. Historical details are given in this book as they contribute to the clarification of the physics under discussion. Here, I offer a thumbnail sketch of historical highlights, and how each of them contributed to the growing body of knowledge about quantum physics. At the same time, I am introducing other concepts of quantum physics which have not been discussed before — in particular, the concept of quantum fields.

Early philosophers and scientists, such as Newton, had questions about the nature of light: are they waves or particles or neither? But it wasn't until 1900 that solid scientific evidence was gathered that began to answer the question. The story goes as follows: when a body of normally black material is heated to high temperatures, it emits light of different colors, much like a glowing metal burner on a cooker. If the target is hot enough, the light becomes bright enough. The spectrum of light can be divided by a prism, and the brightness of each color can be determined by a light detector. When these color-dependent brightnesses were compared to the predictions of classical physics theory, they were found to be defective. The German physicist Max Planck discovered that the problem with classical theory lies in the apparently reasonable assumption that energy could be transferred between hot material and light in any quantity within a continuous range of energies. In an attempt to reach a better agreement between the experiment and the theory, Planck tried to change only one element of the theoretical model. He made the new assumption — radical for his time — that the potential energies exchanged between the material and the light were not constant, but discreet; that is, they occur in stages, as in the stairs.

He concluded that the size of these energy steps is proportional to the frequency correlated with the color of the light being considered. The constant of proportionality is also known as the Planck constant. To the astonishment of the world of physics, this updated model, when mathematically resolved, was in complete harmony with the experimental measurements of the different color brightnesses. Albert Einstein was motivated by Planck's success in proposing a general light hypothesis. He believed that light of a given color could have only a discreet, non-continuous energy content, as would be expected from classical physics. He named these secret quantities of energy, 'light quantum.' Now we call them photons.

Einstein further believed that light quantums are indivisible; that is, they interact as a whole with light absorbing materials. Each photon is either absorbed or not absorbed; it cannot be partially absorbed. He built these ideas into a working theory that correctly predicted how atoms absorb and emit light. His calculations turned out to be, fifty-five years later, the scientific impetus that contributed to the discovery of the laser in 1960. This is another example of the fact that the discoveries of basic science are almost always at the root of the most important technological developments, although the time lag is sometimes very long. Not long after Planck pondered the smooth, rainbow-like spectrum of light emitted by hot objects, other scientists studied the light emitted by gas or vapor containing atoms of only one element (say, neon) when the electrical current passes through it. This is the light we see from fluorescent light bulbs every day. At that time, it was known that the neon atom was made up of a nucleus containing ten protons and normally ten neutrons, surrounded by ten electrons. And it was understood that electrons are 'matter,' because they have mass, unlike photons, which have zero mass. It was thought that electrons were behaving like tiny planets orbiting the nucleus, as if it were a tiny sun. For this model of the atom, classical physics theory predicted that when a neon atom with some excess energy retained by its electrons give up some of that energy, light of any color within a continuous range may be released. But the experimenters found that the light actually shed, consists of only a few well-defined colors, not the predicted smooth rainbow of colors. This was a big mystery, because classical physics theory could not account for this observation.

To make a long story very short, it was realized by 1925 that the flaw in the earlier theory was the assumption that electrons behaved like tiny planets. In other words, electrons should not be treated as small particles, or pieces of matter that take those paths across the nucleus. Louis de Broglie — then a Physics graduate student at the University of Paris — previously postulated that perhaps electrons behave in some way like waves, which was a very non-particle-like view of electrons. Erwin Schrödinger was able to codify this view into a mathematical theory that could precisely predict all the possible wavelike patterns that an electron could create in a given type of atom. He has shown that any wavelike pattern is associated with a particular wave frequency, and therefore, following Planck's concept, with a particular energy. He also discovered that when an electron transitions from a higher energy pattern to a lower energy pattern, the light emits a specific frequency, and therefore a specific color. Schrödinger's equation was able to correctly model all the distinct color variations found in atomic gas light bulbs experiments.

# Chapter 2

## Quantum Physics - The Localization of Manifestation

**Q**uantum physicists speak of electrons, or things that are potential, rather than actual physical entities. And that there are different potentials, simply until someone looks, and then it kind of causes the world to make a decision, so to speak, as to what potential is going to be localized and actualized. All life is basically a limitless quantum field of energy, a sea of infinite possibilities all waiting to happen!

The mind is constructing and manipulating reality. Our thoughts have the capacity to shape our reality. That's how the Law of Attraction functions. We get what we concentrate on most of the time. The observer simply creates reality by observing.

The Mind, no matter what form it appears to contain, holds images. And any picture that is firmly held in the mind, in any form, is bound to come out.

Whenever the mind forms a mental image or an image of anything, it becomes 'ONE' with the Infinite Universal Consciousness, and the image formed is then externalized in the physical world as a unique time-space event. However, in order for the image to manifest, there must be no other conflicting thoughts present to nullify the manifestation power of the image held in the mind.

Another property of the 'quantum' is that it is multi-dimensional. You can now see, scientifically, that our universe is multidimensional, even though our senses are capable of only detecting length, width, height, and time as dimensions. Yet our souls are multidimensional. Listen to your soul, to your feelings.

The physical world is typically made up of thought and energy. Many Quantum Physicists, like Einstein, have shown that all physical matter consists of energy packets that are not bound by space and time.

There are no well-defined boundaries in this energy field. The world is literally your infinite, immortal, unbound body. Science has also shown that the mind has no limits. All minds are 'connected' to a single mind-energy field. You're bigger and more powerful than you think you are.

Whatever you want, you've got it all. It has been said that it will be offered to you even before you ask. Science is slowly coming to terms with it through quantum physics, that this is scientifically true. The infinite knowledge of the formless substance, the potentiality of the quantum level, and our own innate capacity to influence this field, is what gives us the sense of 'having it all.'

You already have all the money beyond your wildest dreams. We are beginning to slowly uncover this on a larger scale, both scientifically and spiritually. You've got it. You may not feel it right now, but you sure have it.

There are two different things to have and learn. A simple way to understand this is, that you have the ability to climb Mount Everest or go Paragliding, but you may not have experienced this part of your ability. There's nothing you need to do to be able to do that; it's already in you. It's been done for you already. All you need to do is experience that skill.

The quantum field can form an infinite number of forms, shapes and experiences. Probably, it's already done that. The page of this book is just one of those things. The terms you're reading are just one of those things, the next idea you're going to have is just one of those things.

But have you ever imagined that you're going to experience these terms through these pages? Your desire to find these words has made them appear in your mouth. In reality, they've always existed. But because of the desire that you and so many others like you have sent across the universe, it has inspired me to give you these answers!

You don't need to predict exactly how things will work out. All you need to do is want, intend, and know it's possible, and it's going to be arranged to come to you. In our lives, we're simply shifting our consciousness to experience aspects of ourselves that we've always had, in a universe that has everything we could possibly ever want to have. There is even something we haven't imagined.

Many physicists working on sub-atomic particles are coming to discover some interesting things about our universe. They have, for example, discovered that two particles separated by space and time can be 'invisibly linked' to each other and act in synchrony. They also found out that the world we live in seems to have been built in such a way that it knows itself.

This seems to have been done by 'cutting' the 'One' whole into at least two halves, one half programmed to see, and the other half to be seen. The one who is conditioned to see is then under the illusion of separation from that which is conditioned to be seen. It's a necessary illusion. But in fact, everything is 'One.'

# Chapter 3

## Quantum Theory - An Overview of the Mystifying Science

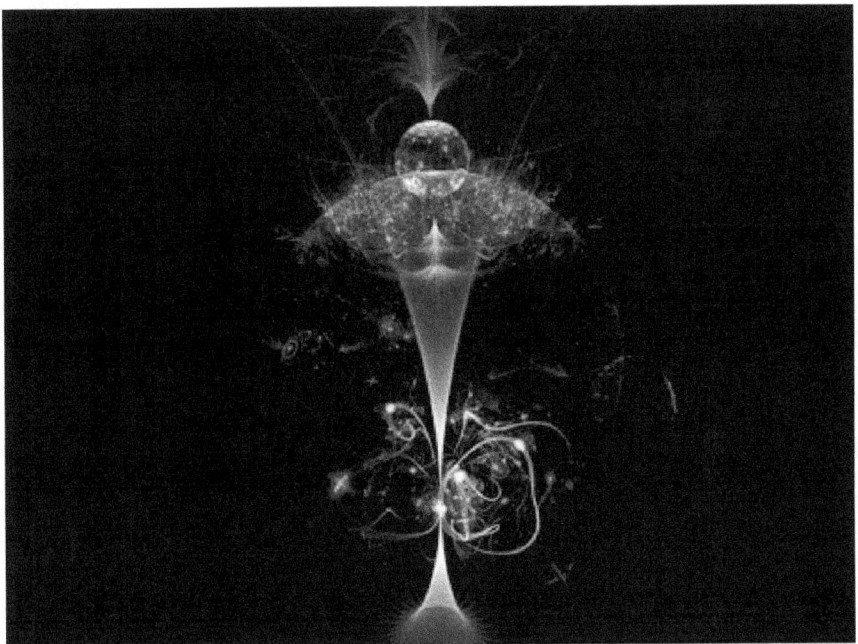

Quantum theory is the most important, fascinating, challenging and even mystifying area of science, and it is considerably more than just weird. It is also the most inspiring concept in the world today. The presumption is that we might well be profoundly mistaken in our assumptions as to what the truth really is.

Originally, the theory was first christened quantum mechanics, considering that it was assumed that there must have been some common laws involved in the activity of atomic particles and quantum energy, similar to the mechanics of the macroscopic subject matter of major planets. The hypothesis attempts to describe the behavior of exceedingly small entities, in general the magnitude of atoms or smaller, in much the same way that Einstein's relativity theory describes the laws of more common entities. This is used in other campaigns, including television and PC's, and also explains nuclear activities in and around the stars.

Quantumists have us living in an infinite number of dimensions furnished in the midst of 'probability waves' and unrecognized 'virtual particles' that flare in and out of existence, but they express verbally that one day we may glide through wormholes within the Universe to look around other cosmoses or to fly backwards in time. In much simpler terms, however, quantum theory is the analysis of the leaps from one energy echelon to another, as it refers to the fabric and behavior of the atoms.

In 1905, Albert Einstein proposed that light was a particle and not a wave, questioning a hundred years of research. He conjectured that not just the energy, but the radiation itself was quantified in the same way. This is the root of Einstein's well-known question that 'God does not play dice.' Einstein certainly could not embrace it as an accomplished science, seeing that quantum mechanics may well, in general, 'only' bestow probabilities on how unique particles react and do not work out certain certainties. For this reason, despite his many novel approximations, Einstein could never let go of the purpose of pre-quantum science to be competent to predict the cosmos like clockwork. Quantum science is not, as Einstein thought, an incomplete science, but, in reality, a rather pragmatic science, inasmuch as it recognizes that in complicated techniques, science will at most give rise to expectations for the reaction of distinct divisions.

Unquestionably, Quantum Theory and Albert Einstein's Theory of Relativity form the basis of modern-day physics, with almost every person conceiving that it is, in fact, the theory of the imperceptible sphere, of tiny particles, and of enormous accelerators. For most people, however, this is a slogan for enigmatic, unfathomable science. This does, however, have a much wider range than just the diminutive sphere and can be adapted to strategies in which several separate parts work with each other while also influencing each other.

### Quantum Physics and the Law of Attraction

It's often hard to understand how the universe works; how you can get what you want, and how sometimes, you just don't seem to get it. The Law of Attraction and Quantum Physics work together to create equilibrium in the universe. It is important to understand both of them, so that you can understand how the Universe works.

First of all, the Law of Attraction – along with Quantum Physics – boils down to a very basic aspect that you need to understand in order to make good use of the Law of Attraction. Like attracts. It is important to remember this fact as you deal with the Law of Attraction, so that you know that you can make the most of the law and what it means.

Basically, when you look at 'Like Attracts,' you look at exactly how it sounds. The way you are, your attitude, your hopes and your dreams, are going to attract similar things to you. The type of energy you bring into the universe is the same kind of energy that attracts you.

Think of the moments when you were angry, upset, and running late. The more upset and frustrated you were about the day, the more late you seemed to be running. The more you dwell on being late, irritated, and angry, the more you see that you actually give yourself more reason to be upset, frustrated, and late. Then think about a good day you've had in your life—a day when everything seems to be going your way. You might be excited and happy, and there seems to be nothing that can bring you down. The more you concentrate on these happy and excited emotions, the more you notice, the more, you're going to be happy and excited.

This is the fundamental idea behind the Law of Attraction—Like Attracts Like. The more you concentrate on good things and positive things, the more the World gives you good things and positive things.

This concept has been around for a long time, but it has only recently become popular, as more and more people begin to understand that the Law of Attraction is actually Quantum Mechanics, a theory of how the universe functions. Quantum Physics teaches that nothing is set, that there are no limits, that everything is vibrating energy. This Energy is under the control of our feelings. It is shaped, formable and moldable. It's different than simply wishing and hoping-it boils down to believing. In order to make the Law of Attraction work for you, you must believe that the Universe will send you the things you really want.

The Law of Attraction could end up being one of the simplest laws you've ever come across. When you fully understand and are able to take advantage of it, you can find that you can have everything you've ever dreamed of.

The Law of Attraction is something that tells a person to draw things to themselves by concentrating on certain things. It has a relationship with Quantum Mechanics, which explains that there is nothing definite and there are no limits. According to Quantum Physics, all is made up of vibrating energy. The Law of Attraction and Quantum Physics are therefore both related and, in fact, interrelated.

According to Quantum Physics and the Law of Attraction, people are the creators of their own universe. The universe is made up of building blocks-not rigid like Newton's classical physics, but fluid and ever-changing, like Quantum Physics.

The Quantum Law of Attraction, therefore, is that because everything is always evolving and fluid—and, in reality, because the Universe is made up of these dynamic and changing energies—everything can be attracted to any person, simply by concentrating. The likelihood that something will happen to someone is very high, as long as it's something they're focused on.

According to Quantum Physics, every person is part of the creation of the universe. That person focuses on issues and attracts them-and, according to the issues that are concentrated on, the items that are brought to each person. Therefore, the World is affected by our feelings. In reality, it's not something that's set—in stone—it's something that's movable and influenced by people's thoughts and what they believe in.

For each person, this means that their dreams may become a reality. All they need to do is focus on the things they want, and the things they've always wanted, and they're going to be able to draw opportunities to themselves much better than they might think they would do. In reality, bringing things to an individual is the only way to obey both Quantum Physics and the Law of Attraction at the same time. Focusing on the things you want and keeping them at the forefront of your mind is the best way to make sure you're motivated to do those things. You'll find that you can do the things you believe in the most easily. It's not always easy to believe that you can have whatever you want—but this is the foundation of the Law of Attraction.

According to Law of Attraction, we attract everything that we constantly focus on. If we think about the relationship between the Law of Attraction and quantum physics, quantum physics explains that nothing in this world is fixed and there are no limitations. Quantum physics also states that all that exists in the universe is vibrating energy.

If you really want to fulfill your dreams and get out of the feeling of being trapped, you need to believe that everything in this universe is energy, and that this energy resides in a state of possibility. You have to allow the rule of attraction to be enforced in order to achieve success. Remember, we are the builders of the universe. According to Newton's classical physics, the universe is made up of discreet building blocks. These blocks are solid, and they cannot be changed.

Quantum physics provides an explanation that there are no separate parts of the universe. All exists in the form of fluid and tends to change from time to time. Physics imagines this world as a deep ocean of energy that keeps coming into existence and disappearing out of this universe.

People living in this world are changing the energy with their thoughts. It is therefore true that one can easily create what he or she wants to achieve. In short, human beings are primarily responsible for the achievement of their goals and the destruction of their desires.

The best thing to understand is that quantum physics has made us the creators of the universe. It's all energy around us.

You must have read Einstein's famous formula. The formula was discovered in 1905 and goes as follows:

$E = mc^2$.

The above formula clearly explains the relation between energy and matter. Energy and matter can be modified quickly. In short, all that exists in this universe is energy, and energy is ever evolving. Our thoughts have a great impact on this energy. Energy can be easily created, molded and formed by our thoughts. We can easily turn the energy of what we think into the energy of what we really want to be.

Quantum physics is also known as the physics of possibility. This theory is contrary to the common idea that the outside world is real and the inside world is fable. It says that whatever happens inside ultimately determines what happens outside the planet. The world in which we live is created by our thoughts.

Nothing is fixed in this world, as mentioned earlier. We need to realize, therefore, that as we concentrate on our thoughts and what we want to draw to ourselves, we can easily get what we want. Still assume that "it can happen" and it will always happen.

The Law of Attraction and its strong connection to quantum physics will allow you to enjoy the success and achievement of your desires. Remember that good things happen to people just because they believe they're going to.

The Law of Attraction and Quantum Physics are closely related. The Law of Attraction notes that through our thoughts and actions, we manifest reality. And not surprisingly, quantum mechanics will explain the Law of Attraction.

The most neglected and misunderstood branch of science at present is quantum physics. Quantum physics looks deeply into the structure of our existence and seeks to explain how the micro influences the macro and to grasp the origin of the Law of Attraction.

Although quantum physics is still not complete, due to the lack of resources to see deep enough to know anything, what has been discovered so far is adequate to understand the Law of Attraction in the world of thought.

One of the most significant discoveries in quantum physics is that matter can function like a particle or a wave. Let me clarify that. A particle is a solid matter—it can only be in one position at a time, so you can still find its spot. However, a wave is not a finite point.

What quantum mechanics has now discovered, through observation, is that when very small particles are fired—called electrons—through two slits, they behave as particles. Each electron picked up a slit, went through it, and hit the back of the screen.

The result of firing hundreds or thousands of these

was a two-slit pattern. However, if the electrons were NOT detected when going through the slits, a broad interference pattern was formed on the back of the screen, which is the effect caused by the wave. In addition, the pattern showed interference from the slits, which further proves that the electrons passed through the slits as waves, not as solid particles.

What does that mean for us, then? Our act of perception, feeling, and emotion has an effect on the environment. When scientists tried to track the electron to predict where it would go, they found that wherever the observer wanted it to end up, it was where it would show up. The consequences of this are equally enormous; our hopes, thoughts and beliefs literally shape the subatomic world around us!

Obviously, the power of our thoughts, emotions, desires, and values to affect change, and construct reality is just what the Law of Attraction tells us. Now that you have some scientific background, you might be able to put aside your current beliefs and try it out. If by any chance you were told that you could have everything you wanted, would that at least be worth a try? Suspend your disbelief, and be astounded.

# Chapter 4

## Quantum Physics and You

We just entered the age of the Aquarius. This means that our Solar System has moved into a completely different place in the Galaxy. We, the people living on planet Earth, have never been here before. What we can expect from this new space that we occupy remains to be seen. At the beginning of any great cycle, there are always harbingers, clues and broad stroke rules for what we can anticipate. We've only been living in this place in space for fifty years. We've still got another 1950 years to go. What this new "Day" has revealed to us is already very important in the path that we should take.

First of all, the "Pisces Age" we've just left is over. Although many of the theories, structures and forms of the Age of Pisces continue to be strong and alive, their plug has been pulled. They are no longer linked to the power of the Galactic Sun. The Galactic Sun is now driving the Aquarius Age.

The Age of the Pisces was known as the Age of "I Believe." It served its purpose because humanity needed to build its belief system from the 'Authorities.'

The Age of Aquarius is the Age of "I Know." We will be able to get all the knowledge we need to grow and evolve directly from the source. There's no need for a middle man.

This is made possible by one of the first gifts given to us by the "Day of Aquarius." It is the "Age of Mind," the gift of Quantum Physics is for the mind. These laws will lay down guidelines for us to use for the next 1950 years.

They simply tell us that there is an infinite ocean of thought, an intelligent energy called the Quantum Ocean. In reality, it's the mind of God.

All that has ever been, is or is going to exist there. There is no time, past, present or future. Only the one NOW. There's no room here. No Width, no Length, no Depth, only the HERE.

In reality, the Quantum World, the Mind of God, is an infinite point called the HERE-NOW. And how we can seemingly live within this infinite point as well as outside this infinite point on Planet Earth is still a mystery to our finite minds. It will become glaring to us as we pass through the 2000-year period. For now, we need to use the Laws of Quantum Physics and the concept of the Quantum Ocean, the Mind of God, to recreate a new reality for ourselves and for mankind in general.

Another theory is that the Quantum Ocean, the Mind of Nature, is open to our "Thoughts." Thoughts are things! We need to really do more than just think about the Quantum Ocean, the Mind of God, we need to learn how to get there. How to be right there. We need to learn how to coexist 24/7, both psychologically in the Quantum Ocean and physically on the planet Earth. We need to understand how to use the Laws of Quantum Physics in the Quantum Ocean to restore our life and our world.

**Quantum Physics and Your Health**

The Laws of Quantum Physics posited that everything is energy. There is really only one basic force, and it resides in the Quantum Ocean. Scientists are banding back and forth theories and postulates about what is the basic building block of Creation. They're calling it "Quanta." They're talking about particles and waves, about possibilities and odds. Words, Words, Words! Definitions are less abstract than the object itself.

The term "apple" in a piece of paper is not a tree. You

can't eat the word "steak" written on a piece of paper. The energy that occurs in the Quantum Ocean and 'blinks in and out of' our physical reality is the basic building block of Creation.

Where did this come from? God, or more precisely, the Quantum Ocean, is the Mind of God. It's the one thing that materialistic scientists leave out of their wit, talking and braying about Quantum Physics.

They will not come up with a correct answer until they place the Creator, God, in their equation. As to who we are, where we came from and where we are going, the 'Big Bang Theory' concept does not have the correct answer to these questions. It doesn't answer the question, "Who put the energy in the ball and started the momentum in the first place?" The Creator, God, did.

When we know more about the Laws of Quantum Physics and the depth and breadth of the Quantum Universe, the true reality of the Creator, God, will be too immense for us to comprehend with our finite minds. We need to explore the Quantum Ocean, the Mind of God, and work individually with the energies of the Creator.

All of the Laws of Quantum Physics point to the intelligence behind Creation. The Quantum Ocean is known to be an infinite Ocean of Thought, Intelligence, Energy that responds to our thoughts and emotions. It's the intellect behind the development.

Since there is a Creator, therefore, there is also a design of the Maker. This concept is in the "mind" of the Creator. And because the Mind of the Creator is the Quantum Ocean, there is a strategy inside the Quantum Ocean.

Everything that has ever been, is, or will ever be, is contained in the mind of the Creator, God, the Quantum Ocean. It is in the form of the Divine Blueprints, the Divine Archetypes. These patterns are set up in such a way that when they 'blink out' into the Loom of our physical reality, they take definite forms.

There are Divine Blueprints for the Perfect Animals, Men, Women, Birds, Planets, Galaxies, and the Universe. When these Divine Plans initially 'blinked out' of God's Mind, they could be conceived as the Big Bang. It was fast. Each of us is an individual soul within the mind of God. We're 'blinking out' and 'in,' Birth, Death, just as the universes are. There's a Divine timing scheme for all of this. This is what the Bible's reference to "There's a Season" means for everything!

In order to achieve better health, we must aspire to balance our present energy configuration (mind and body) with the original energy configuration of our Soul. We should learn to enter the Quantum Ocean, search for and conform to our individual Divine Blueprint. We've got to burn off all the wrong "Dross" we picked up on our journeys. The Quantum Ocean is the place where the blueprints of perfect health exist. Time to enter and retrieve the Quantum Ocean, the Mind of God.

## Quantum Physics and Its Connection to your Self-Esteem

No matter what your beliefs are, the universe has secrets that cannot be explained and behaves in a way that is far from haphazard. It was Albert Einstein who said, "All this does not happen at random," and science is now moving more and more towards the idea that the universe is behaving intelligently and that there is a source of energy that shapes it.

The latest understanding is that the universe is made up of Dark Matter, Dark Energy (they don't know what it is) and Sub -atomic Particles in the form of matter and stuff that fills the void or gaps between matter, i.e. Space. We're the Energy!

Sub-atomic particles are not particles as such, they are particles of energy which, when observed, turn into solid objects, that is the matter we see in our universe and in our everyday life. Albert Einstein's formulae, $e = mc^2$ tells us that e (energy) = m (mass) x c (light speed) is squared.

This tells us that All Mass is equal to Energy. ALL MASS is equal to energy, not just sun or gas or oil or uranium. ALL MASS = Energy. It shows that everything in the universe, galaxies, stars, planets, moons, gases, and the US is energy. We 're not made up of "we are stuff" energy (there's enough energy in the human body to fuel a small town for two weeks).

Such sub-atomic particles do not just sit there in mass to form matter, they vibrate, float, and morph in and out of the body in waves, forming and re-creating in a dance that is directly linked to changes in observation. They have intellect and fly through time and space without hindrance. They exist in the past, present and future, and at the present time they will respond and build on the basis of information gathered from all these dimensions.

They don't judge what's right or wrong, and they'll form whatever you ask for, but they'll give you warnings and lessons to guide you along your path if that's what you need. But, of course, it's up to you whether or not you accept this and change your desire to reflect on the messages you're receiving. It's called Free Will!

It is said that everything that can exist already does, and that it is through our observation of it (through thought) that creates it in our experience.

Take Thomas Edison, who performed thousands of experiments to discover the light globe, to give you an example. Using the Law of Attraction and by focusing on the end result he wanted, Edison knew all he had to do was persevere and he would eventually find the right formula to invent the light globe. The light globe always existed as a possibility, but only became a reality when Edison finally observed it as it should have been after thousands of experiments.

By the laws of attraction (The Secret) he knew that he would eventually achieve his goal. Why it took thousands of experiments to make this happen would have a lot to do with the thought processes that went through his mind at the time. It could have been his ego, his doubts or his limited beliefs, but one thing is certain that, through persistence, he eventually achieved his goal, and the light globe became a reality as soon as he finally saw the globe as it is. That is, when he was finally vibrating at the same frequency as his goal.

When we concentrate on thinking, experiencing, or believing in what we want from the sub-atomic particles that make up everything in the world, including space, we begin to work together and build it in our lives.

What we want is always a possibility, and the role of the world is to give it to us and turn back to satisfy us for whatever we want. It wants to give you what you want, as a matter of fact, that's its purpose. Forces far beyond our understanding come into play, and the arrangement of events begin to move into a position that will lead to circumstances that will give us all we want.

At some certain point in our life, we all face problems. Sadly, it seems like most people don't know how to let them go and cling on to them, even though the root of their issues might be long gone. It is only a matter of creating a condition where it can be observed to become a reality in our experience. It's the same thing to have what we want or to be who we want to be. It already exists as a possibility, but it is not in our experience until we observe or vibrate at the same frequency or, in other words, believe it to be so. If we study it, believing it to be so, and giving it our attention, the world starts to form it into our experience.

We're starting the process with our thought, our attention, and as the time buffer required to build what we want is going on, it continues to change to our reality. Similarly, in order for us to make, we need to think about it first. Thought is like a blueprint that the universe uses to create our reality. Therefore, if we want to improve our lives or get what we want, we need to think or send thought to the world so that it can construct our new truth.

It needs a new collection of blue prints to give us what we want!

All in all, all that can actually exist does; this means that the possibilities of your future, as you want it to be, actually exist. It's just a matter of experiencing your life as you want it to be, or vibrating at the same frequency as you want it to be, and the world will, over time, manifest it in your perception and become your reality.

Low self-esteem and loss of trust are negative emotions and the mechanism of sabotaging constructive manifestation. Anyone who suffers from these afflictions focuses on what is wrong with life, rather than what is right or good in life. They cannot understand why their life causes so much suffering and believe that, feeling bad for themselves, the world must feel sorry for them. They spend all their time in self-pity in vain, hoping that the universe will change their circumstances and bring about a miraculous cure for their unhappiness, but they never respond.

The less they respond, the more they feel sorry for themselves and the process continues to spiral downwards. It is only you who can change your circumstances and your unhappiness. As we have shown, the world gives you what you want and gives it in abundance. It will build the world around you in your head, and it will continue to do so until it has been told differently.

If you suffer from lack of confidence or low self-esteem, what you think is that which manifests your suffering in your life. We have discovered that the universe produces and assembles what you want, using your thoughts as blueprints. The clearer and more focused your thoughts, the better and faster you can build what you want. You draw the blueprints using your positive thoughts as instructions, and the universe begins to build as soon as the pen is put on paper, so to speak.

It foresees what you want, drawing on your past, present and future experiences, and begins to assemble them entirely on the basis of the messages you are sending at the moment. We're all living the past... By that I mean, we are living the result of our thoughts, our words and our actions of the past (in other words, the many past yesterdays). If your thoughts, words and actions were loving, happy, positive, specific and clear yesterday, then that's the life you're going to lead today. If your thoughts and actions yesterday were full of disappointment, sorrow, remorse, rage or dissatisfaction or lack, then that's the life you're going to live today.

You can't change your yesterday or your today, but you can change the tomorrow you want, TODAY!

The universe does not distinguish between negative or positive energy of thought. If you focus your attention on what you want with clarity and trust and hope, you will receive it. If you center your attention on what you don't want with honesty and conviction, you'll obtain it as well. It's very important that you understand that. Your ideas, whether positive or negative, are used as directions (blueprints) to build and/or assemble what you want, whether you want it or not.

Low self-esteem and loss of trust are negative emotions and the mechanism of sabotaging constructive manifestation. Anyone who suffers from these afflictions focuses on what is wrong with life, rather than what is right or good in life. They cannot understand why their life causes so much suffering and believe that, feeling bad for themselves, the world must feel sorry for them.

They spend all their time in self-pity in futile hopes that the world will change their circumstances and bring about a magical solution for their unhappiness. They begin to believe, "This is my lot in life, and I might as well settle for less." They say to themselves, "I'm not good enough," and in doing so, nothing ever changes and the cycle continues.

Only you can change your circumstances and your unhappiness, and it begins from within. As we have seen in the universe, using sub-atomic particles gathers for you what you want, whether you want it consciously or not, and gives it to you in abundance. It will build the world around you in your head, and it will continue to do so until it has been told differently.

If you suffer from a lack of trust or low self-esteem, ask yourself, "What kind of thoughts do I think would make this misery a reality?" Be aware of your thoughts and use your feelings as indicators. If you feel bad, it tells you not to go there. If you feel good, it tells you this is the path you should take. Trust yourself! The universe has given you the tools to guide you.

Don't stress that if you fail or it doesn't happen as quickly as you want, the universe will give you exactly what you need to move forward, even if it looks like a step backwards. The natural process of life is more life, and it's against the forces of nature to reduce life. Every time you think you're going backwards, you're actually moving forward, all you need to do is learn from your mistake.

Look at your life as if you're in the audience rather than the main character of a movie. Look at the mistakes and look for ways to fix them. As the word says they're only missing—takes in a movie, they may be irritating at the moment, but all we need to do is go back to the scene and move on to the next one. Your life is a movie in your mind, make it a happy one.

# Chapter 5

## The Building Blocks of Matter and Wave-Particle Duality

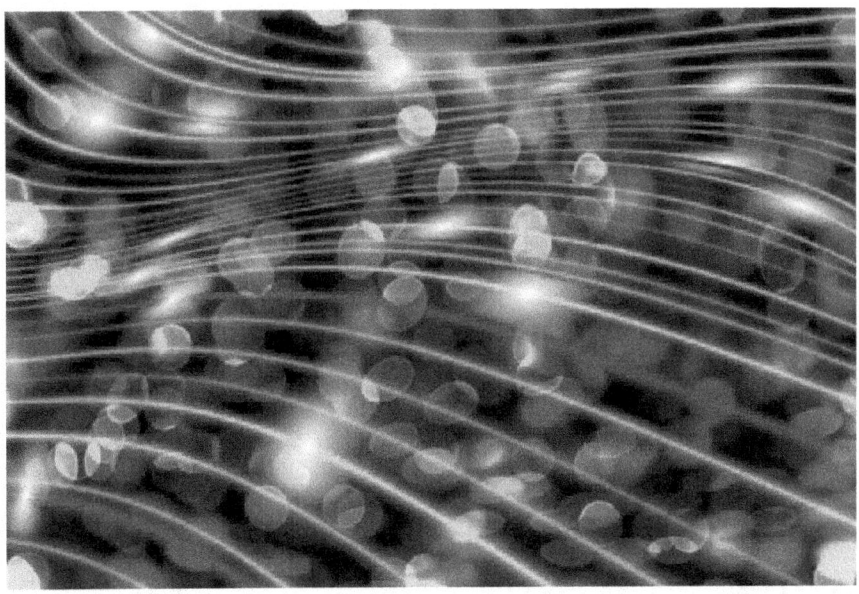

One of the big all-time questions is, 'What are things made of?' Most of the answers have come down on the side of having a set of basic building blocks to make it all. This was initially studied by philosophers, then by alchemists and then by chemists. It took the physicists to figure this out!

I'm very sure you've heard of one of the first models we know about. Everything was made of a combination of four elements—earth, air, fire and water. This had the advantage of having only a few building blocks and had connections between properties and contents.

The next major model was the chemist Mendeleev's periodic table. All matter consisted of atoms, with one type of atom per element (e.g. iron, oxygen). Over 100 elements have been identified—a huge number to be simple building blocks!

Take a piece of gold with you. Keep cutting it down into smaller pieces. Every lump is always going to be gold. Finally, it was assumed that you might get to the stage where you couldn't cut it any more. The ancient Greek word 'atomos' (meaning 'uncut') has been used to refer to such small lumps of an element. The word means 'indivisible'—the small lumps were thought to be fundamental.

However, physicists have discovered that electrons have come from within atoms. That meant that the atoms had to be made of something else. The hunt was going on. Rutherford studied the Plum Pudding System of JJ Thompson in the early 20th century. This held that the negative electrons had been held in a positive 'batter.' He set some of his students, Geiger and Marsden, to test it.

Some things we're happy to call stuff; tables, chairs, squirrels. Some of the things we're happy to call waves; sound, water ripples, Mexicans. Some things, however, have led to long arguments about what they are. Light is a very important example. Back at the end of the 17th century, Newton assumed that light was made up of tiny particles (he called them corpuscles) and so was matter. Huygens assumed that the light was waves. At that time, Huygens won the debate with a series of experiments showing how light actually behaved like a wave; it could spread through a gap (diffract). The matter seemed resolved before we started to be able to study the sub-atomic world.

In 1905, Einstein published a paper on a dark phenomenon called 'photoelectric effect.' It was observed that when light shone on some of the metals charged, they lost their charge. It wasn't all metals, however, or all light forms. Zinc, for example, maintains its charge when white light shines on it but loses it when ultraviolet light (the kind used in tanning booths) shines on it. This could not be clarified if the light was a wave. Einstein realized that this was evidence that Newton was correct and that the light was actually made of particles. He named them the 'quantum of light' or the 'photons.' A branch of physics called quantum mechanics has been born.

But hang on, Huygens' experiments have shown that light is a wave. Those experiments are still working today. So, what's going on; can't they both be right? Well, they really are. It appears that light is both a wave and a particle. It only behaves differently depending on the circumstances. Sound familiar to you? Yeah, it's just like mass-energy and space-time. In this case, the concept is called 'wave-particle duality,' and we say that light is made of wavicles (from WAV-PARTICLES). So, if we strongly believed that light was a particle, but it turned out to be a wave, what about the things that we strongly believed were particles?

In 1906, JJ Thompson was awarded the Nobel Prize for proving that electrons were particles. He did this by showing that they had quantified the mass and the charge-they came in fixed lumps instead of being able to have any amount. In 1937, his friend, George Thompson, was awarded the Nobel Prize for proving that electrons were waves. Nowadays, we know that both of them are wavicles, and both Thompsons were right.

**Quantum Physics - The Discovery that Scientifically Demolished Materialism**

Quantum physics shows that sub-atomic particles appear and disappear spontaneously in a vacuum. Interpreting this observation as matter may arise at the quantum level, this is a property pertaining to matter, some physicists try to explain the origin of matter from non-existence during the creation of the universe as a property pertaining to matter and to present it as part of the laws of nature. In this model, our universe is represented as a larger subatomic particle.

However, this syllogism is definitely out of the question and cannot, in any case, explain how the universe came into being. William Lane Craig, author of The Big Bang: Theism and Theism, discusses why:

Quantum mechanical vacuum spawning material particle is far from the usual idea of "vacuum" (meaning nothing). Rather, quantum vacuum is a sea of constantly forming and dissolving particles that borrow energy from the vacuum for their brief existence. This isn't "nothing," and therefore, material particles don't come out of nothing.

And, in quantum physics, matter doesn't work when it wasn't before. What happens is that ambient energy somehow becomes matter and, just as suddenly, energy disappears again. In short, as stated, there is no state of life from nothingness.

According to Isaac Newton, light was the flow of a substance known as a corpuscle. The foundation of conventional Newtonian physics-which was believed before quantum physics was discovered-was that light consisted entirely of a collection of particles. Although, James Clerk Maxwell, a 19th century physicist, suggested that light represented wave action. Quantum theory has reconciled the biggest controversy in physics.

In 1905, Albert Einstein claimed that light had quantum or small packets of energy. Such energy packets have been called photons. Although described as particles, it could be observed that photons behaved in the wave motion proposed by Maxwell in the 1860s. Light was therefore a transitional phenomenon between wave and particle, a state of affairs that showed a major contradiction in Newtonian physics.

Immediately after Einstein, Max Planck, a prominent German physicist, studied light and astonished the whole scientific world by determining that it was both a wave and a particle. As a result of this idea, which he proposed under the name of quantum theory, energy was disseminated in the form of interrupted and discreet packets, rather than being straight and constant.

In a quantum event, light showed both particle-like and wave-like properties. The particle, known as the photon, has been followed by a wave in space. In other words, light moved through space like a wave, but behaved as an active particle when faced with an obstacle. To put it another way, it took the form of energy until it met an obstacle at which point it assumed the shape of particles, as if it were made up of tiny material bodies reminiscent of grains of sand.

This theory was further expanded after Planck, by scientists such as Albert Einstein, Niels Bohr, Louis de Broglie, Erwin Schrödinger, Werner Heisenberg, Paul Adrian Maurice Dirac and Wolfgang Pauli. Every one of them was awarded the Nobel Prize for their discoveries.

About this new discovery concerning the nature of light, Amit Goswami says:

When light is seen as wave, it seems to be capable of being in two (or more) places at the same time as when it passes through the slits of the umbrella and produces a diffraction pattern; when we capture it on a photographic film, however, it appears discreetly, spot by spot, like a particle beam. So the light has to be both a wave and a particle. Paradoxical, isn't that? One of the bulwarks of old physics is at stake: a clear description in the language. The concept of objectivity is also at stake: does the nature of light—what light depend on—affect how we perceive it?

Scientists have no longer assumed that matter consists of inanimate, random particles. Quantum physics had no materialistic meaning, since the nature of matter was non-material stuff. While Einstein, Arthur Holly and Philipp Lenard Compton were investigating the particle structure of light, Louis de Broglie began to look at its wave structure.

De Broglie's observation was extraordinary: he found in his experiments that sub-atomic particles often exhibited wave-like properties. Particles like the electron and proton also had wavelengths. In other words, within the atom — which materialism represented as absolute matter — there were non-material waves of energy, contrary to materialistic belief. Just like light, these tiny particles inside the atom at times acted like waves and displayed the properties of the particles in others. In contrast to materialistic standards, absolute matter in the atom could be observed at certain times, but vanished at others.

This major discovery has shown that what we believe the real world is, in reality, shadows. Matter had finally departed from the world of physics and was going in the direction of metaphysics.

Now we know how the electrons and the light are behaving. But what do we call it? If we say that they behave like particles, we give the wrong impression; even if we claim that they behave like waves. They act in their own inimitable way, which could theoretically be called a quantum mechanical manner. They're acting in a way that's like nothing you've seen before. The atom does not behave like a weight that hangs on a spring and oscillates. Nor does it serve as a miniature representation of the solar system with little planets in orbits. Nor does it seem to be like a mist or a fog of some kind, circling the nucleus. It's like nothing you've ever seen before.

There is at least one simplification. In this respect, electrons behave in exactly the same way as photons; they are both dirty, but in exactly the same way. So how they behave takes a lot of imagination to appreciate, because we're going to describe something that's different from anything you know about. Nobody knows how that could be.

To sum up, quantum physicists say the objective world is an illusion. All the most renowned physicists of the 1920s, from Paul Dirac to Niles Bohr, and from Albert Einstein to Werner Heisenberg, sought to explain these results from quantum experiments. Eventually, one group of physicists reached an agreement known as the Copenhagen Interpretation of Quantum Mechanics at the Fifth Solvay Conference on Physics held in Brussels in 1927-Bohr, Max Born, Paul Dirac, Werner Heisenberg and Wolfgang Pauli. This name was taken from the place of work of the group leader, Bohr, who suggested that the physical reality proposed by quantum theory was the information we have about the system and the estimates we make on the basis of that information. In his opinion, these assumptions made in our minds had nothing to do with outer reality.

In short, our inner world had nothing to do with the outer real world that had been the main subject of physicists' interest from Aristotle to the present day. Physicists abandoned their old ideas about this view and agreed that quantum understanding represented only our knowledge of the physical system. The material world that we can perceive, exists only as information in our brains. In other words, we can never have direct experience of matter in the outside world.

Perhaps the most important, and most insidious assumption that we absorb in our childhood is that of the material world of objects that exist outside — independent of subjects who are observers. Circumstantial evidence exists in favor of this assumption. When we look at the sky, for example, we consider the sky where we expect it in its classically determined trajectory. We imagine, of course, that the moon is still there in space-time, even though we're not looking at it. Quantum Physics is saying no. When we're not looking at it, the moon's probability of wave expands, but by a tiny number. When we look, the wave falls immediately, because the wave must not be in space-time. It makes more sense to adapt the idealistic metaphysical assumption: there is no object in space-time without a conscious subject looking at it. This applies, of course, to our perceptual world. The presence of the Moon is visible in the outside world, of course. Yet as we look at it, the only thing we really experience is our own understanding of the Moon.

The function of observation in quantum physics cannot be overemphasized. In Classical Physics [Newtonian

Physics], the observed structures have a mind-independent nature that observes and probes them. In quantum mechanics, however, only by means of an act of observation can a physical quantity have an actual value. What makes things happen is not more things. But it's ideas, concepts, information that make things up.

Following the most sensitive and fascinating experiments that the human mind could have devised over the course of 80 years, there are now no views opposed to quantum physics that have been decisively and scientifically proven. No challenge can even be raised to the conclusions drawn by the experiments carried out. Quantum theory has been evaluated in hundreds of different forms by scientists. It has received the Nobel Prize for a variety of scientists and continues to do so.

Matter, the utmost fundamental concept of Newtonian physics, once viewed unconditionally as absolute truth, has been eliminated. Materialists, the supporters of old belief that matter was the only and definitive building block of existence, were really confused by the lack of substance suggested by quantum physics. They now have to explain every laws of physics in the field of metaphysics.

The shock inflicted on materialists at the beginning of the 20th century was far greater than can be described in these pages. But quantum physicists, Bryce DeWitt and Neill Graham have defined it as follows:

No advancement of modern science has had a deeper influence on human thought than the emergence of quantum theory. Wrenched out of centuries-old patterns of thought, the physicists of a generation ago found themselves forced to embrace new metaphysics. The distress caused by this reorientation continues to this day. Essentially, physicists have experienced a significant loss; their grip on reality.

# Chapter 6

## Quantum Possibilities and Waves

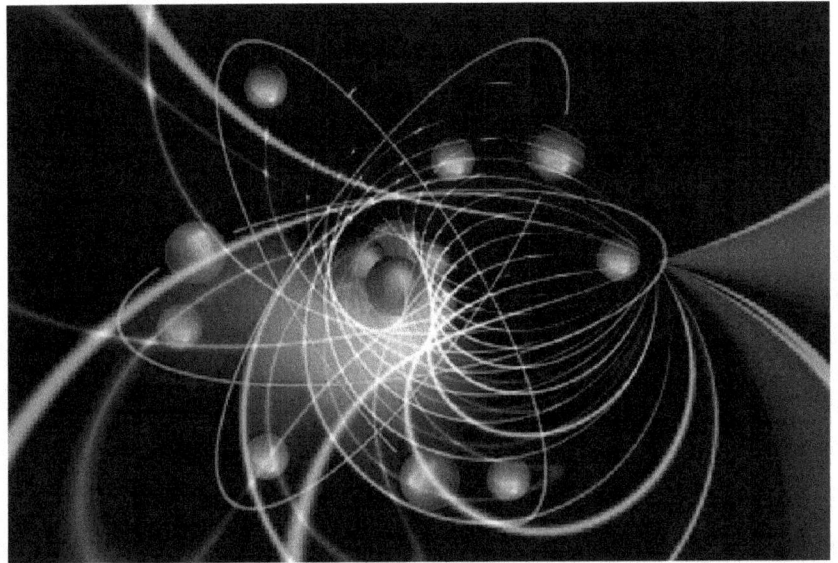

In quantum theory, the probabilities are determined by considering all the possibilities that may be involved in a given process. If there are intermediate possibilities involved in the process, quantum theory tells us how to combine them in order to find the resulting possibilities. The probabilities for a particular measurement outcome can be calculated from these resulting possibilities. Quantum possibilities for objects such as electrons act in some respect as waves do.

But the electron is a single particle, so what's 'waving?' Typically, we think of a wave as a movement on some extended physical medium, such as sound waves through the air or ripples on the surface of the lake. The same physical definition of a wave cannot apply to a single electron, which, if detected, is found to be at a level, not to be distributed within a certain area. However, the wave concept can be applied to single electrons because it correctly explains how the quantum possibilities that correspond to different measurement results change over time and vary across space.

## What are the waves?

Waves are orchestrated patterns that travel across space. If a rock is thrown into a pond, it creates ripples on the surface of the water that travel away from a rock-hit location. The ripple pattern travels across the lake, while each water molecule oscillates around its own fixed positions. The molecules rise and fall, causing the adjacent molecules to also rise and fall, time-delayed slightly from the motion of the surrounding molecules. This synchronized motion of water molecules contributes to the transfer of energy and momentum along the surface. The duck that floats some distance from the wave source (the rock entry point) will be influenced by this energy and momentum, and will oscillate up and down. Note however, that this flow of energy and momentum does not involve the water actually flowing between the wave source and the location where its effects are felt (the duck). An example of a wave is shown in FIGURE 1.0 as a moving pattern. Maximum wave height locations are called 'crests' and minimum wave height locations are 'troughs.' The pattern between two adjacent crests (or two troughs) is called a full cycle. The distance between the two adjacent crests is called the length of the full cycle (also called the 'wavelength'). The pattern or wave tends to move forward, but there are no objects moving in the direction the wave moves.

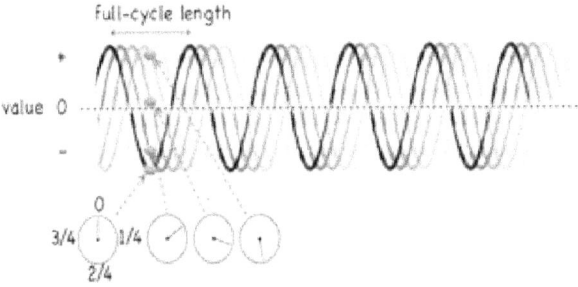

Figure 1.0 A wave is shown to have a magnitude (wave height for a water wave) that oscillates positively and negatively in a normal manner. The gap between the crests is the length of the complete span. The time elapsed during one complete cycle of oscillation is the full cycle time.

The point (or duck) on the wave oscillates up and down once at a characteristic frequency called the full-cycle frequency (also called the 'phase' oscillation). The figure shows a clock that holds time while a specific point on the wave oscillates up and down. The rate at which the clock hand rotates around the clock face depends on the springiness of the wave medium, such as water or air, and its density. When the internal clock of the wave reads zero, the wave pattern is located as shown by the darkest curved line in FIGURE 1.0. As time passes and the clock hand rotates, the wave pattern shifts smoothly to the right, as seen by the series of lighter curved lines. Every time the clock turns around once, the wave travels a distance equal to the duration of the complete cycle.

The speed at which the wave passes is equal to the amount of time divided. This is wave speed: wave velocity = full-cycle length/full-cycle time. This relationship between full-cycle length, full-cycle time and wave speed can be visualized in a cartoon-like way with

a fictitious ladder, as in FIGURE 1.1.

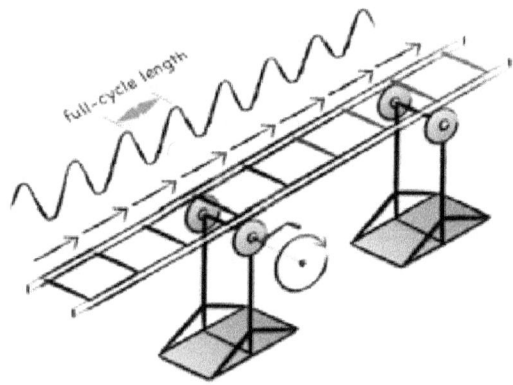

Figure 1.1. Mechanical description of the relationship between full-cycle time, full-cycle length and wave speed.

The distance between the ladder rungs is the full-cycle length of the wave. The ladder rests on a mechanism designed to move it using a crank-and-wheel arrangement. The wheels attached to the crankshaft grasp the bottom edge of the ladder by friction, so that when the crank is rotated, the ladder is pushed in the direction shown by the arrows. The diameter of the wheels is such that if you turn the crank around completely one time, the ladder travels at a speed equal to the distance between the rungs of the ladder. Therefore, if you turn the crank at a steady rate, the ladder is continuously propelled forward at a steady speed. For example, if you turn the crank once every second and the rungs are separated by one foot, the speed of the ladder is one foot per second.

## What is Wave Interference?

If two rocks are thrown together into a pond, each creates waves in the form of ripples that move away from the location where the rock hits. When the duck floats on the shallow surface, the impact of both waves can be felt. In certain points in the pool, the behavior of the two waves reinforce each other, creating a broad up-and-down oscillation (the duck has a crazy ride). This reinforcing effect is called positive interference. But, at certain places in the bay, the behavior of the two waves may be reversed, causing no up-and-down oscillation. (The duck is still there.) This canceling effect is called destructive interference.

# Chapter 7

## Application: Quantum Computing

**Is information physical?**

Computers are devices that process information. Computer scientist and physicist, Rolf Landauer, argued that knowledge is a part of the physical world. He elaborated this as follows: information is not a disembodied abstract entity; it is always linked to physical representation. It is represented by engraving on a stone tablet, a [magnetic] turn, a [electrical] charge, a hole in a punched card, a mark on paper, or some other equivalents. This links the handling of information with all the possibilities and limitations of our actual physical world, its laws of physics and its storage of available parts. If "information is physical," as Landauer has said, then it would seem necessary to treat it mechanically. In other words, the physical means by which information is stored and interpreted by computers should be considered using quantum theory. It helps to understand computation in general before addressing quantum computers.

## What is a Computer?

A computer is a machine that receives and stores information input, processes the information according to a programmable sequence of steps, and produces the resulting output of information. The term 'computer' was used for the first time in the 1600s to refer to people who perform calculations or computations, and now refers to computers that compute. Computing machines can be divided roughly into four types:

1. Computing devices for computational classical physics. These machines use moving parts, including levers and gears, to perform computing. Usually, they are not programmable, but always perform the same operation, such as adding numbers. An example is the 1905 Burroughs Adding Machine.

2. Electromechanical classical mechanics fully programmable computing devices. These machines operate using electronically controlled moving parts. They process information stored as digital bits represented by the locations of a large number of

electromechanical switches.

The first such machine was built in 1941 by Konrad Zuse in wartime Germany. In theory, their programmability allows them to solve any problem that can be found and overcome by using algebra. These were the first 'universal' computers in this context.

3. All-electronic, hybrid quantum – classical – physics computers. These fully programmable, universal computing machines have no moving mechanical parts and work using electronic circuits. The first to be constructed was the ENIAC, engineered by John Mauchly and J. Presper Eckert, University of Pennsylvania, 1946. The physical principles that describe the motion of electrons in these circuits are rooted in quantum physics. But, given that there are no superposition states or entangled states involving electrons in different circuit components (capacitors, transistors, etc.), classical physics adequately describes the manner in which electrons represent information. Therefore, we call these machines—essentially any computer in operation today—'classic computers.'

4. Quantum computers. If ever built successfully, these devices would operate according to the principles of quantum physics. Knowledge will be expressed by the quantum states of individual electrons or other elementary quantum artifacts, and there will be entangled states involving electrons in various circuit components. These computers are expected to be able to solve those kinds of problems much faster than any

modern classical computer can do.

## How do Computers Work?

Computers store and manipulate information using an alphabet binary language consisting of only two symbols: 0 and 1. Each 1 or 0 symbol is referred to as a bit, short for a binary digit because it can take one of two possible values. A page of text, such as the one you read, is represented in a computer file as a long string of numbers. Every letter is represented by a binary code. For example, 'A' becomes 01000001, 'B' becomes 01000001, and so on.

In a typical computer, each bit is represented by the number of electrons stored in a small device called a capacitor. We can think of a capacitor as a box that holds a certain number of electrons, sort of like a bulk grain bin in a food store that holds a certain amount of rice. Each capacitor is called a memory cell. For example, such a capacitor could have a maximum capacity of 1,000 electrons. If the capacitor is full or almost full of electrons, we say it represents a bit of a value of 1. If the capacitor is empty or almost empty, we say it represents a bit value of 0. The capacitor is not allowed to be half-filled, and the circuitry is designed to ensure that this does not happen. Through grouping together eight capacitors, each of which is either full or empty, any eight-bit number — e.g. 01110011 — can be interpreted.

The role of the machine circuitry is to empty or fill various capacitors according to a set of rules called a program. Eventually, the action of filling and emptying the capacitors manages to perform the desired calculation — say, to add two 8-bit numbers. In a computer, the actions are performed by tiny components of computer circuitry called logic gates. Logic gate is made of silicon and other elements arranged in a way that either blocks or passes electrical charge, depending on its electrical environment. Logic gate inputs are bit values, represented by a full capacitor (a 1) or an empty capacitor (a 0). (The word 'gate' is associated with the fact that something goes into it and something comes out of it.)

## How small can a single logic gate be?

In the first all-electronic computers, such as the ENIAC, built in the 1940s, a single logic gate was a vacuum tube similar to the amplifier tubes still used today in vintage-style electric guitar amplifiers. Each tube has at least the size of your thumb. By 1970, the microcircuit revolution was able to reduce the size of each gate to about one-hundredth of a millimeter. When things get much smaller than this, it's best to measure the length of a unit called a nanometer, which is equal to one millionth of a millimeter. The size of the gate in 1970 was 10,000 nanometres. On the other hand, a single silicon atom, which is the main atomic element in computer circuitry, is around 0.2 nanometer in thickness. By 2012, single gates in typical computers had been reduced enough so that they could be spaced apart by as little as 22 nanometers — that is, only about a hundred atoms apart. The actual working area of the gate was less than 2.2 nanometres, or 10 atoms in thickness. This small size allows you to place a few billion memory locations and gates in an area the size of your thumbnail.

Having gate sizes much smaller than those dimensions leads to both a curse and a blessing. We leave the domain of many-atomic physics and enter the domain of single-atomic physics. There are now variations between the classical physics principles that well explain the average behavior of many atoms and the quantum physics principles that are required when dealing with single atoms. We reach a random action domain that doesn't sound good if we're trying to get a well-regulated system to do our numerical bidding. In reality, a group of scientists led by Michelle Simmons, director of the Center for Quantum Computation and Communication at the University of New South Wales, Australia, constructed a gate consisting of a single phosphorus atom embedded in a silicon crystal tube. This is the smallest gate ever to be designed. This gate only functions properly if cooled to an extremely low temperature: – 459 degrees Fahrenheit (– 273 degrees Celsius). If the material is not at least as cold, the random (thermal) motion of the silicone atoms in the crystal decreases the confinement of the electron psi wave, which may leak out of the channel into which it is intended to be confined. For day-to-day desktop computers, which, after all, have to operate at room temperature, this leakage prevents such single-atom gates from being the basis of technology that everyone can use. On the other hand, these experiments demonstrate that computers can, at least in theory, be built on the atomic scale, where quantum physics rules.

Can we create computers that use fundamentally quantum behavior?

Given that physics defines the ultimate behavior and efficiency of information transfer, storage and processing, it is reasonable to ask how quantum physics plays a role in information technology. Because electronic computers rely on the behavior of electrons, and communication systems rely on the behavior of photons — both elementary particles — it is not surprising that the performance of information technology is ultimately determined by quantum physics. But here is a subtlety. Computer technologies currently in use do not involve quantum superposition states to represent information. They use states that can be considered classical states of physical things — namely, groups of electrons.

The big question is: can we create computers that use quantum mechanical states to enhance our ability to solve real-world problems? If these computers were ever built, they would be able to bypass certain forms of data encryption methods much faster than any computer that is operating today. This would revolutionize the field of privacy and confidentiality for computers and the Internet. The encryption key that might take thousands of years to crack using a conventional computer could only take minutes on a quantum computer.

## What is a Qubit?

The word bit is used to refer to both the abstract, disembodied mathematical concept of information and the physical entity that embodies the information. It is clear in classical physics that a 'physical bit' carries one 'abstract bit' of knowledge. There is a very simple one-to-one relation between the state of the physical bit and the value of the abstract bit, 0 or 1. We may also use individual quantum artifacts, such as an electron or a photon, to incarnate a portion. In this case, the elementary physical entity is called a qubit, short for 'quantum bit.' A qubit has two different quantum states, such as the H and V polarization of the photon, or the upper path and the lower path of the electron. When measured, the results represent a bit value of 0 or 1. But remember that we can select different polarization measurement schemes — say, H/V or D/A. The results may then be random, with the probability of observing possible outcomes depending on which measurement scheme we selected. In this case, there is no one-to-one relation between the state of the physical qubit and the value of some conceptual abstract bit. The concepts of quantum physics suggest significant variations between the behavior of classical bits and qubits. Classical bits can be copied as many times as we want, without any degradation of the information; qubits cannot be copied or cloned even once, although they can be teleported. The state of the classical bit, 0 or 1, can be determined by a single measurement; the quantum state of the single qubit cannot be determined by any sequence of measurements.

## What physical principles differentiate classical and quantum computers apart?

There are large differences between the types of gates used in classical computers and the gates that need to be used in quantum computers. Classical gates perform operations that are not reversible; understanding the output does not tell you what the inputs are. On the other hand, if a quantum gate is to operate properly with qubits, it must be reversible. That is, you need to be able to determine the input states by understanding the output states. This requirement arises because any quantum gate operation must be a unit process. Recall that we use the word 'unitary' to refer to physical processes or behaviors that cannot be divided into individual steps, each with definite, observable outcomes. Our main example was an electron (or photon) passing from source to final location in a situation where two separate paths are possible. We pointed out that if there is no permanent trace left by the passage of the electron that indicates that it has taken a particular identifiable path, it is wrong to say that it has actually taken one path or the other. It's also not correct to say that both directions have been taken. The whole process of leaving one place and arriving at another must be seen as an undivided, complete process — that is, a unitary operation. These processes are reversible.

## What logic gates would quantum computers use?

Since the early 1990s, scientists have been theorizing how a universally programmable computer based on quantum superposition and entanglement could be constructed, and for which kinds of problems it would be ideally suited. Neither is it a simple issue, nor has it been completely resolved to date. On the other hand, considerable progress has already been made and the chances seem reasonable to decent that such a machine will become a reality within, say, ten or twenty years. A quantum computer takes qubits as inputs, performs a series of gate operations on them according to a program designed by the programmer, and outputs the modified qubits. The number of output qubits must be equal to the number of input qubits for the whole process to be unitary. As is the case with classical computers, there are different choices that the designer will make for the set of gates to be used in a quantum computer. I am focusing here on a set that is similar to the logic of {XOR, AND} described above for classical computers. For quantum computers, I follow the logic I call {QXOR, QR}. Using two quantum gates, called 'quantum XOR' and 'quantum ROTATE,' a universal quantum computer can be built, at least in principle. What are these two kinds of gates doing?

First, note that we give the two possible qubit states the names 0 and 1 and they are represented, for example, by the H-polarized and V-polarized states of a single photon. The general operating principles of a quantum computer are independent of how we choose to represent the qubits of physical objects.

The quantum XOR gate or the QXOR gate is shown in FIGURE 2.1. In the QXOR gate, the B qubit moves unchanged through the gate (from left to right) rather than being discarded as is the case in the classical example. This makes the QXOR gate both reversible and unitary. That is, outputs are uniquely linked to inputs.

A good way to think about the QXOR gate is to say that qubit B controls what happens to qubit A, as indicated by the arrow pointing from B to A. If qubit B is in state (0), then qubit A passes through unchanged, as in parts I and (ii) of FIGURE 2.1. But if B is in state (1), then the state of qubit A is changed from (0) to (1) or from (1) to (0) as in FIGURE 2.1 sections (iii) and (iv).

Figure 2.1 I Quantum XOR gate. Qubit B controls the process of qubit A modification. The Qubits are moving from left to right.

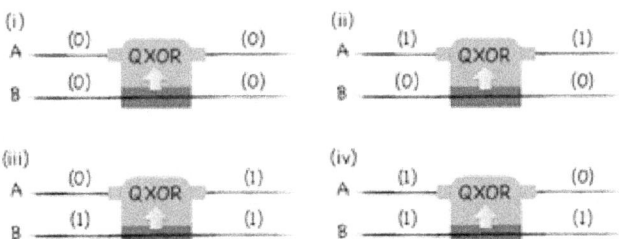

The second type of logic gate required is the quantum ROTATE gate, or QR gate, shown in FIGURE 2.3. This gate has one input and one output, and is both reversible and unitary. If the input state is (0), the output state is the same superposition of the (0) and (1) states, with the state arrow pointing in the diagonal direction. If the input state is (1), the output state is again the same superposition of the (0) and (1) states, but with the state arrow pointing in the antidiagonal direction. Both 'a' and 'b' arrow components have a value of 0.707 (i.e. one-half square root). According to Born's Law, this implies the likelihood of obtaining either the outcome, (0) or (1), of measuring each equal to 0.5 or fifty per cent.

Figure 2.3 The quantum-ROTATE gate produces a superposition of (0) and (1) qubit states. The arrow-rotation diagrams are similar to the polarization state arrow diagrams.

These two potential output states are similar to diagonal and anti-diagonal polarization. In the event that the qubits are represented by a photon polarization, the QR gate is easily implemented using a special crystal that rotates the polarization state arrow by forty-five degrees in the counter-clockwise direction. Quantum computing theorists have shown that by combining sequences of

QXOR gates and QR gates, any qubit-based computation can be done.

## How Would Quantum Computers Operate?

The classical machine works by defining input data, in the form of bit selection, and sending those bits to the processor where the gates function sequentially according to the program, and then reading the bit values at the output.

The quantum computer works by defining input data in

the form of a set of qubits each with its quantum state specified, sending those qubits to the quantum processor where the gates act on them according to the program, and then measuring the qubits at the output. The main difference between the classical case and the quantum case is that there can only be overlap and entanglement in the quantum case. These quantum states can only occur in the circuit if the overall operation of the gates together is a unitary process. The process is unitary only if there is no way, even in principle, for a person to know any of the individual qubit values (0 or 1) in the inner part of the circuit.

## Which physics and chemistry problems can quantum computers solve?

The history of quantum computing did not begin with computer science, but with physics. In 1981, Richard Feynman, one of the most inventive theoretical physicists, found out that the fundamental equation of quantum theory — Schrödinger's equation — cannot be solved efficiently by ordinary computers. Schrödinger's equation plays a role in quantum theory like Newton's motion laws in classical physics theory. The difference is that, while Newton's laws explain how classic objects behave in terms of definite and perfectly predictable outcomes, Schrödinger's equation describes how quantum states shift in time. Again, note that quantum states are not in one-to-one correspondence with the measurement results, but show only the potential for results. The fact that Schrödinger's equation cannot be easily overcome by using ordinary computers is a big obstacle for the advancement of science. We have a fundamental equation that we need to solve to predict the probabilities of experimental outcomes; but, in the case of sufficiently complicated situations involving many quantum objects, we can't solve it! We simply don't know exactly what the theory predicts, so we can't make full use of it to advance science, engineering, and medical research. We can't develop better quantum-based drugs because it's not possible to solve Schrödinger's equation for large molecules. Of course, scientists have many ways of finding approximate solutions to Schrödinger's equation, which is very helpful, but we don't have accurate solutions that might contain welcome surprises. Feynman assumed that a new type of computer he called a quantum computer would be able to solve Schrödinger's equation efficiently. Since Feynman pointed this out, a lot of work has gone into trying to build a computer like this. Such a computer would operate according to quantum principles rather than the classical principles of physics, as ordinary computers do. Unlike most computer science problems and math problems, problems involving Schrödinger's equation can easily be turned into algorithms that can be performed on a quantum computer. That's because Schrödinger's equation is the fundamental equation of quantum theory!

For example, Schrödinger's equation allows us to calculate the energy and shape of the psi wave for each possible quantum state of the electron within the atom. Molecules, such as the all-important DNA molecule, are made of atoms arranged in ways that create their structure and enable them to perform their functions, such as encoding and propagating a person's genetics. Because DNA molecules contain so many quantum particles — electrons, protons, and neutrons — using classical computers, it is impossible to precisely understand and predict the structure and function of Schrödinger's equation. To see why Schrödinger's equation for a DNA molecule using a classical computer is so difficult to solve, consider a simple example. Let's assume that the molecule contains a minimum of five hundred electrons. In fact, this is a relatively small molecule compared to DNA. To represent the quantum state of these five hundred electrons using the bit state of the computer, it is necessary to represent all possible entangled states of the five hundred electrons.

Each of these possible states represents a quantum possibility for a particular combination of results that could be observed if measurements were made on all electrons. To keep it simple, let's say that each electron could be in one of two states, labeled 0 or 1—that is, it can be thought of as a qubit. There are four possible variations if there are two electrons: 00, 01, 10, and 11. There are eight potential variations if there are three electrons: 000, 001, 010, 011, 100, 101, 110 and 111. There are 2500 or about 10150 possible combinations of states for five hundred electrons that need to be considered. This is much greater than the total number of elementary particles in the whole universe! Each combination must be represented by a number in the memory of the computer, but it is impossible to store all of these numbers in any computer smaller than the entire universe. A solution could be to split all the combinations into smaller groups and process each group individually by moving the numbers into and out of the memory of the computer. But the time needed to make all this move would probably take longer than the life of the universe. This example illustrates Feynman's main point: as the size of the quantum problem to be solved grows larger, the size of the computer needed to solve it grows even faster — exponentially, in fact.

# Chapter 8

## Why are Quantum Computers so Difficult to Make?

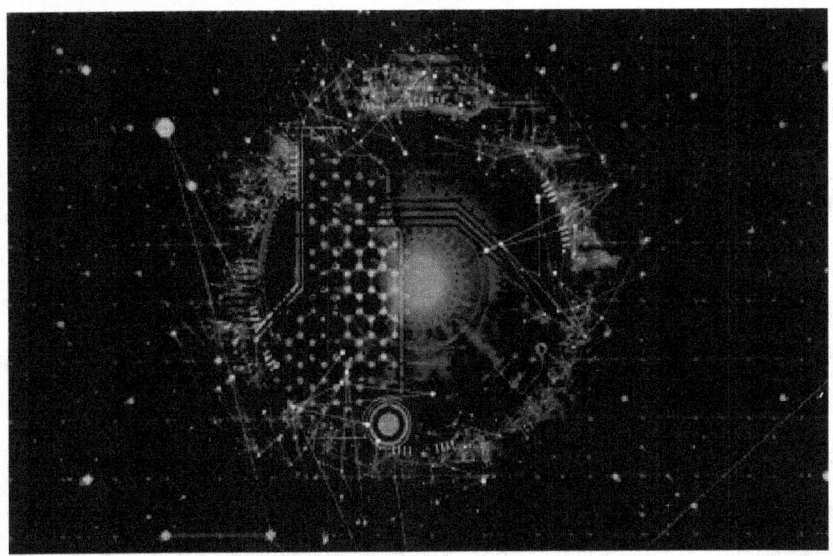

I f not very well managed, quantum computers should be much more error prone than traditional computers. This intense sensitivity is due to the complexity of holding all qubits in the right superposition state during the entire computation. Recall, for example, that the qubit superposition states (0)+(1) and (0)+(−1) are separate states. Yes, they give the same probabilities for a 0 or 1 result if the measurement is performed, but they are physically quite different. Find a qubit described by a single photon to see this. The polarization state (H) + (V) is a diagonally polarized (D) state, while (H) + (− V) is an anti-diagonally polarized (A) state. Their state arrows are perpendicular, and so completely different. These countries are very delicate. For example, the unintended addition of a very small timing difference between the H and V components of the diagonal state can transform it into an antidiagonal. This would completely disrupt the intended quantum computation.

Errors in computing caused by unwanted disturbances or 'noise' would prevent a quantum computer from giving the correct answers if there was no way to predict, detect, and correct these errors as the computation proceeds. Fortunately, by using quantum theory, physicists have discovered ways to correct such errors in a working quantum processor. The idea is to include some extra qubits in the input, the purpose of which is to keep track of any unwanted errors. These extra qubits are intertwined in a special, known way with the qubits that we care about — the ones that do the computing. Then, by measuring extra qubits, without disturbing the qubits that we care about, we can detect an error that might have occurred. Such error detection is reminiscent, but not identical, of the detection of errors in the quantum encryption key distribution setup. If an error has been found, it can be corrected before the calculation continues. Unfortunately, adding more and more qubits, which also need to be controlled almost perfectly, adds a great deal to the complexity of the quantum computer, making it very difficult to build these computers.

## What are the Prospects for Building Quantum Computers?

This of course, is a kind of difficult question to answer since the goal is a moving target. It is fair to assume that there is no universal quantum computer in operation. Many small-scale demonstration experiments have been carried out successfully, which seem to show that the physics on which the promise of quantum computing rests is indeed strong. These demonstrations seem to show that building such a computer is now 'only' a matter of ingenuity and extremely challenging engineering, rather than a matter of fundamental physics. Scalable means that if you can build a quantum processor that contains, say, one hundred qubits, then it would only be twice as hard to build one with two hundred qubits, only three times as hard to build one with three hundred qubits, and so on. If you can build a quantum processor that contains, say, one hundred qubits, then it would be twice as hard to build one with two hundred qubits. That is, you don't want the difficulty of building, or the size, or the amount of resources needed, to grow exponentially with the growing size of the problem you're trying to compute. This would defeat the whole aim of building quantum computers to solve the exponential scaling problem.

## What are the Promising Techniques to Building Quantum Computers?

While several methods are being studied, the three most promising platforms for constructing quantum computers are perhaps superconducting electronic circuits, isolated individual atomic ions trapped magnetically in a vacuum chamber, and isolated individual phosphorus atoms embedded in silicon crystals. Although the international race to 'get there' first, is worthy of a book of its own. A nucleus containing protons and neutrons is at the core of a phosphorus atom. The nucleus functions like a small permanent bar magnet, with north and south poles. Such a magnet may be positioned with a north pole pointing up or down. A qubit is represented by the orientation of this magnet: UP is 1 and DOWN is 0. The orientation of the nucleus magnet can be controlled briefly by applying a magnetic field, the force of which causes the magnet to rotate in a different direction. This enables the quantum QR gate operation needed as part of the quantum computer operation described above. To construct a quantum computer, many phosphorus atoms are required to represent several qubits, and QXOR gate operations involving pairs of qubits need to be performed. The many phosphorus atoms are arranged in a pattern like a chessboard, with a phosphorus atom at the center of each square having dimensions of 30 nanometres by 30 nanometres. Recall that the size of a single atom is about 0.2 nanometer, so these are very small squares! Normally, the qubits contained in the internal magnet orientation of each phosphorus atom do not influence each other. This is 'quiet time,' when the qubit values are simply stored. The whole silicone crystal must be cooled to an extremely low temperature: − 391 degrees Fahrenheit or − 196 degrees Celsius. This prevents the internal magnets from being buffeted by the excessive jiggling of the silicone atoms that make up the crystal, which could lead to the nucleus magnets being accidentally rotated in the wrong direction. As we discussed earlier, this would lead to errors in the qubit states and would entail the use of error correction methods. If researchers want to operate a QXOR gate between two nearby phosphorus atoms, they activate each by passing one electron from nearby wires to each of the atoms. For reasons relevant to atomic physics, the internal magnet in each atom is much stronger, and they tend to influence each other. (If you've ever kept two magnets close to each other, you know the stronger one tends to force and rotate the other one.)

The arrangement of the two phosphorus atoms results in the action of QXOR as follows: the magnet in one of the phosphorus atoms, called qubit B, regulates what happens to the magnet in the other phosphorus atom, called qubit A. That is, if B equals 0 (DOWN magnet), then qubit A remains unchanged. However, if B equals 1 (Magnet UP), then the magnet qubit A is pushed and flipped from 0 to 1 (DOWN to UP) or from 1 to 0 (UP to DOWN).

## Future Direction in Quantum Science

Quantum scientists need to better understand the non-intuitive aspects of quantum phenomena and their description by quantum theory in order to make further progress. They need to develop new technologies for building devices that rely on such phenomena for their operation. Quantum physicists are working to understand more deeply the non-local correlations that can occur when measuring quantum-entangled objects.

Quantum theory can be used to describe the observed non-local correlations using the concept of entangled states of two objects. Such an entanglement can occur even though two objects are separated by a wide distance in space. The quantum description makes it clear that what happens to one particle does not directly affect the state of the other particle; yet the correlations still occur in a manner that defies the description of classical physics. Physicists would like to know how long such correlations occur. Although quantum theory describes these correlations perfectly, it does not say how they come to be. Is there a 'backdoor' channel that somehow transmits the correlations without violating the cause-and-effect nature of things that the Theory of Relativity seems to require?

What happens to non-local correlations when the two quantum objects are near or inside a black hole, where space-time is strongly distorted? In such regions, the concepts of time and space require us to rethink what we mean by 'local.' Questions like this could lead to breakthroughs in understanding the nature and early history of the universe itself. Quantum technologists are working to increase their skills in the design and construction of devices needed to advance the three main areas of research and development: quantum communications, quantum sensors and quantum computing.

Technological breakthroughs are needed, for example, in the following areas: the first includes light sources that reliably generate single photons at known, controllable times. The challenge is to overcome the randomness of quantum results in the production of such photons. The second necessary breakthrough involves small, portable nuclear interferometers for inertial sensing. These devices depend on the ability to cool a small, confined cloud of atoms to temperatures near absolute zero. Third, methods for constructing a 'scalable' quantum computer—that is, a scheme in which doubling the size of computer memory and processors requires only twice the cost and space (not an exponential increase in memory). An example of this is the need for improved techniques for placing single atoms at known locations in a solid material such as silicon or a magnetic trap in an evacuated chamber and the means to manipulate and test their quantum states. All of these, and more, are topics of intense, ongoing study. We still do not know how far quantum technology can be pushed to create alternatives to classical physics-based technologies with enhanced capabilities. Will efficient, functioning quantum computers be designed successfully and, if so, what tasks will be used most productively? Can quantum-based sensor technology mature and be implemented in a wide range of applications? Or will the nature of these instruments and their extreme sensitivity to minor disruptions make them uneconomic?

New technologies are not likely to replace existing ones, but will increase or supplement them, and each will be used where it is best suited. This prediction is similar to the idea that both classical physics theory and quantum theory have their position and their role in explaining physical systems. We use the one that is best suited to solve a particular problem at hand. It is most important to know what you don't know, to make progress in science. It is essential to ask the right questions. You may still find yourself puzzled by some aspects of quantum phenomena and their theoretical description.

Above all, we have discovered that Existence is probabilistic — that is, that some events occur in an essentially random manner. For example, if an electron is excited to be in a state of high energy in an atom, it will decay to lower energy, releasing a photon. It may decay at any time after excitement, and there is simply no way to predict precisely when this will happen. On the other hand, causality is still maintained; a specific event cannot occur unless the necessary prior events have occurred and the necessary prior conditions have been met. In view of certain prior events and conditions, there is more than one possible future, each with its own probability of occurrence, but each has to be consistent with the ideas of cause and effect.

We currently have the existing quantum theory, which covers almost everything we know about: classical and quantum. And we still do not have a completely good picture of what quantum theory is trying to teach us about Nature, or at least not one that is agreed upon by the vast majority of physicists. There is a community of physicists trying to get to grips with the underlying reasons why the observed phenomena are best described by quantum theory and the form it takes in terms of state arrows, superposition, unitary processes, and so on. While agreeing that the theory we now have 'works', allows for the design and construction of new quantum technologies, these physicists hope that exploring the underlying basis of quantum theory at its utmost fundamental, even philosophical level will lead to further breakthroughs.

# Conclusion

The main thing you should know about Quantum Physics is that when the electrons are not watched, they behave like waves showing wave interference patterns. When the electrons are watched, they turn into actual particles of matter and show the patterns of a single line that you would expect to shoot particles through slits. What does that mean? Here's how we're going to interpret it. Get ready to…

There's nothing separate. In reality, the entire universe, including you and I, are pure energy. It is through our thoughts that we transform this energy into what we see as reality. Remember, in the video, the waves became particles when they were watched. Our thoughts are creating the world in which we live. The paradigm shift is that we have always assumed that the outside world is truer than the inner world. But the opposite is true of that. What's going on inside of us determines what's going to happen outside of us. With our thoughts, we create our world.

Now, just ask yourself this question. What does it mean for energy to play the piano, to play the guitar, to sing, to speak French or to set up a computer? In our three-dimensional world, we use our conscious mind to see energy as its physical counterpart. This makes us play the piano, play the guitar, etc. We are the force that is every single thing in the universe. Which means that it cannot be partial to any single individual. If you can do that with one guy, you can.

When waves of energy are detected, they break down into particles. This is how the whole world is made. We build our universe by collapsing waves of possibility into our existence with our conscious mind. Probability is the calculation of how likely an occurrence is to occur; a number that represents the ratio of probable cases to the total number of potential events.

We basically have an infinite number of probability wave functions that can actually collapse. What you want to break down into reality is up to you. You literally have millions of alternatives, some of which may be piano playing, guitar playing, singing, speaking French, or computer programming.

So what we want to do is increase the likelihood of bringing the things we want into our reality and decrease the likelihood of bringing the things we don't want into our reality.

### How are we going to do that?

We send out vibrations through our thoughts, or to be

more precise, through the emotions that our thoughts generate. Different emotions are vibrating at different frequencies. Positive emotions are vibrating at higher frequencies. Negative feelings are vibrating at lower frequencies. To increase the probability of collapsing the probability wave function if you want to become a reality, you need to focus on what you want while feeling positive emotions.

www.ingramcontent.com/pod-product-compliance
Lightning Source LLC
Chambersburg PA
CBHW050247220526
45465CB00002B/587